SCIENTIFIC AMERICAN EDUCATIONAL PUBLISHING

GET MOVING!

10 FUN

EXPERIMENTS ABOUT SPORTS AND MOVEMENT

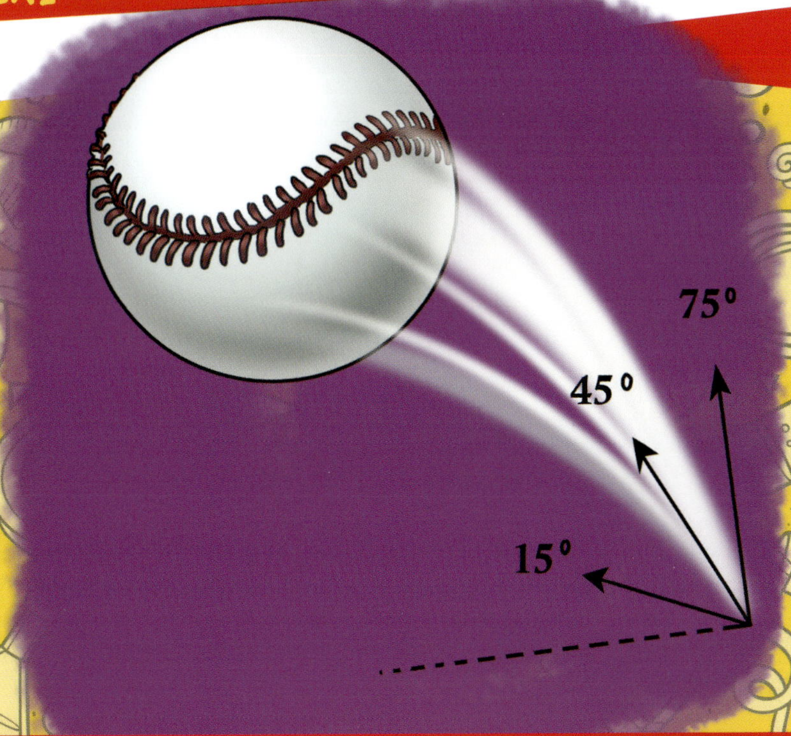

BRING SCIENCE HOME

Published in 2024 by Scientific American Educational Publishing in association with
The Rosen Publishing Group 29 East 21st Street, New York, NY 10010

Contains material from Scientific American , a division of Springer Nature America, Inc., reprinted by permission as well as original material from The Rosen Publishing Group .

First Edition

Editor: Kelly Meredith
Designer: Rachel Rising

Activity on page 5 by Science Buddies (September 6, 2012); page 9 by Science Buddies (May 17, 2012); page 15 by Science Buddies/Ben Finio (March 21, 2019); page 21 by Science Buddies/Sabine De Brabandere (February 22, 2018); page 27 by Science Buddies/Sabine De Brabandere (April 13, 2017); page 33 by Science Buddies (May 28, 2015); page 39 by Science Buddies (May 1, 2014); page 45 by Science Buddies (July 11, 2013); page 51 by Science Buddies (March 6, 2014); page 56 by Daisy Yuhas (May 24, 2012).

All illustrations by Continuum Content Solutions

Photo Credits: pp. 3, 5, 7-9, 12, 18, 24, 30, 33, 36, 42, 45, 49, 54, 55, 59 cve iv/Shutterstock.com; pp. 4, 5, 9, 15, 21, 27, 33, 39, 45, 51, 56 Anna Frajtova/Shutterstock.com.

Cataloging-in-Publication Data

Names: Scientific American, Inc.
Title: Get moving!: 10 fun experiments about sports and movement / edited by the Scientific American Editors.
Description: New York : Scientific American Educational Publishing, 2024. | Series: Bring science home | Includes glossary and index.
Identifiers: ISBN 9781725349063 (pbk.) | ISBN 9781725349070 (library bound) | ISBN 9781725349087 (ebook)
Subjects: LCSH: Force and energy--Experiments--Juvenile literature. | Motion--Experiments--Juvenile literature. | Sports--Experiments--Juvenile literature. | Physics--Experiments--Juvenile literature. | Science projects--Juvenile literature.
Classification: LCC QC73.4 .G48 2024 | DDC 531'.6--dc23

Manufactured in the United States of America

CPSIA Compliance Information: Batch #SACS24. For further information contact Rosen Publishing, New York, New York at 1-800-237-9932.

CONTENTS

⚛ THESE ACTIVITIES INCLUDE SCIENCE FAIR PROJECT IDEAS.

INTRODUCTION

Sports are a great way to have some fun while getting some exercise. But did you know there's a lot of science involved too? In these experiments, you'll see why skis are the length they are, learn if there's a method behind carnival games, test reaction times, and more—all while you see how science and sports go together.

Projects marked with include a section called Science Fair Project Ideas. These ideas can help you develop your own original science fair project. Science fair judges tend to reward creative thought and imagination, and it helps if you are really interested in your project. You will also need to follow the scientific method. See page 61 for more information about that.

Skydiving Science

Does the Size of a Parachute Matter?

HOW DOES A PARACHUTE WORK? DO BIGGER PARACHUTES WORK BETTER THAN SMALLER PARACHUTES? FIND OUT IN THIS SCIENCE PROJECT IF THE SIZE OF THE PARACHUTE MATTERS.

Have you ever wondered how a parachute works—or which design features are most important in slowing someone's descent? Parachutes come in many different shapes and sizes, but they work based on the same general principles. In this activity, you will test differently sized parachutes to see how changes in the size of the parachute affect flight. What do you think will work better: a bigger parachute or a smaller one?

PROJECT TIME

30 to 60 minutes

KEY CONCEPTS

Aerodynamics
Drag
Flight
Forces

BACKGROUND

In the pastime of skydiving, a person jumps out of an airplane from a very high altitude, falls through the air, and releases a parachute to help them slow their way down and land safely on the ground. How does the parachute break the free fall so well?

As the skydiver is falling, the force of gravity is pulling the person and their parachute toward Earth. The force of gravity can make an object fall very fast! The parachute slows the skydiver down because it causes air resistance, or drag force. The air pushes the parachute back up and creates a force opposite to the force of gravity. As the skydiver falls, these "push and pull" forces are nearly in balance.

MATERIALS

- Heavy-duty garbage bag
- Scissors
- Ruler
- String
- Four pennies
- Tape
- A safe, high surface, about 6.5 feet (2 m) from the ground. A good place may be a secure balcony, deck, or playground platform.
- Stopwatch (accurate to at least 0.1 seconds)

PREPARATION

- Cut open the garbage bag to make a flat sheet of plastic.
- Cut two squares out of the garbage bag. Make one 8 inches by 8 inches (20 cm by 20 cm) and one 20 inches by 20 inches (50 cm by 50 cm).
- Tie a knot in each of the four corners of each square.
- Cut the string into eight pieces that are each 16 inches (40 cm) long.

- Tie one end of each piece of string around each of the knots, positioning the string right above the knot.
- For each square, hold the center of the square in one hand and pull all of the strings with the other hand to collect them. Tie the free end of the strings together with an overhand knot.
- Securely tape two pennies to the end of the strings on each square. *What do you think the purpose of the pennies is?* Hint: you can try letting the squares float to the ground without the pennies first and see what happens.
- Your parachutes are now ready to test!

PROCEDURE

- Bring a stopwatch and the parachutes to the safe, high surface you found that is about 6.5 feet (2 m) from the ground.
- Release the smaller parachute from high above the ground and time how long it takes for it to fall to the ground. Try this parachute two more times, releasing it from the same height each time. *About how long did it take to fall on average?*
- Release the larger parachute from the same height and time how long it takes for it to fall to the ground. Try this parachute two more times. *About how long did it take the larger parachute to fall?*
- *Which parachute took longer to fall to the ground? Why do you think this is?*

EXTRA

Make some more parachutes out of differently sized squares from the garbage bag. Test out your new parachutes. *Do you see a clear trend between the size of the parachutes and how long it takes them to fall?*

SCIENCE FAIR IDEA

In this activity, you tested one variable—the surface area of the parachute—but there are a lot of other variables that affect how well a parachute works. Try this activity again but this time vary the material that the parachute is made out of (you could try nylon, cotton, tissue paper, etcetera), the shape of the parachute, the length of the string, the weight of the string, the load (by increasing or decreasing the number of pennies), or the height at which the parachute is dropped. *How does changing one of these other variables affect how well the parachute works?*

OBSERVATIONS AND RESULTS

Did the larger parachute take longer to fall to the ground than the smaller parachute did?

How large a parachute is (in other words, the parachute's surface area) affects its air resistance, or drag force. The larger the parachute, the greater the drag force. In the case of these parachutes, the drag force is opposite to the force of gravity, so the drag force slows the parachutes down as they fall. Consequently, the larger parachute, with its greater drag force, takes longer to reach the ground than the smaller parachute. Although the force of gravity is greater on the larger, slightly heavier parachute than the smaller, lighter one, the relative increase in the drag force on the larger parachute is greater than the increase in the force of gravity.

CLEANUP

Remove the pennies from the parachutes. Throw away the garbage bag pieces. Return all other materials to where you found them.

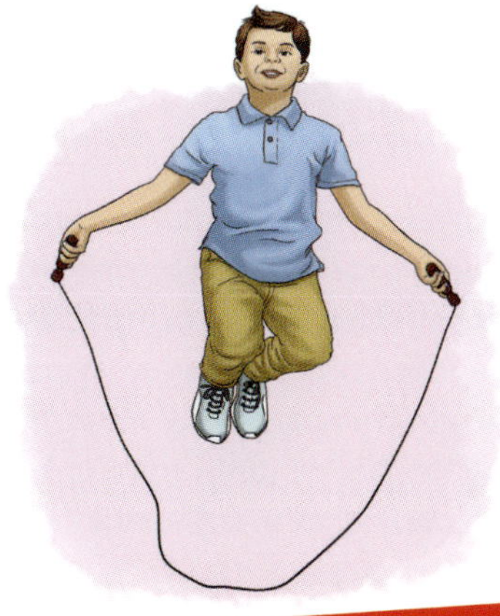

Skipping Science

An Experiment in Jump Rope Lengths

WANT TO JUMP FASTER? THINK THE LENGTH OF YOUR JUMP ROPE IS A FACTOR? IN THIS EXPERIMENT, YOU'LL LEARN IF IT IS!

Have you ever wondered how you could do jump rope faster? The U.S. jump rope record for the greatest number of jumps in one minute is 367! That's more than six jumps a second! How close do you think you can get to that number? What are some of the factors that will help you jump faster? One is the length of the jump rope!

PROJECT TIME

30 minutes

KEY CONCEPTS

Coordination
Endurance
Speed

BACKGROUND

Jumping rope is great exercise. Professional boxers do it to improve their coordination and endurance. Plus, it can be a lot of fun! That's easy to see when watching students jumping rope together, often competing for speed or doing tricks. And there are official competitions for speed jumping. What would you do to improve your speed? The length of the jump rope can change how many jumps a person can make in a minute, but what length of jump rope is ideal? The longer the jump rope, the more time it takes to turn it in a full circle. The shorter the jump rope, the faster it turns, but because the circle is smaller the jumper might have to jump higher to get over the rope. That might slow the person down or cause them to make a mistake.

MATERIALS

- A 10-foot (3-m) jump rope
- Two volunteers who know how to jump rope (can include yourself)
- Stopwatch or watch with a second hand
- An open, flat space large enough for jumping rope

PREPARATION

- Choose an open, flat space that has enough room for jumping rope. (Make sure nothing is nearby that you might bump into while you jump.)

PROCEDURE

- Fold the jump rope in half to find the midway point. Stand on this point with both feet. Put a handle in each hand and pull the handles straight up along your sides.

- Shorten the rope by moving the handles halfway between your belly button and armpits, then tie knots in the rope just below the handles. This is the shortest jump rope length you'll test.

- Have the volunteer time you while you jump rope for one minute. As you jump, you or the person timing you should count how many jumps you do. When one minute is up, the person timing you should tell you to stop. If you "mess up" within the minute do not stop, but continue timing, jumping rope, and counting the jumps. For example, if after 10 successful jumps, the rope hits your foot and you have to restart, the counter should count the next successful jump as number 11. *How many successful jumps did you do in one minute using the shortest jump rope length?*

- Readjust the rope length so that the tips of the handles are now just barely brushing your armpits. This is the medium length you'll test.

- Again, have the volunteer time you while you jump rope for one minute and count how many jumps you do in that minute. *How many successful jumps did you do a minute using the medium length? Was it greater or less than the number of jumps you made using the shortest rope length?*

- Readjust the rope so that the tips of the handles just barely brush your chin. This is the longest length you'll test.

- Again, have the volunteer time you while you jump rope for one minute and count how many jumps you do during that period. *How many successful jumps did you do per minute using the longest rope length? Was it greater than or less than the number of jumps you made using the shortest and medium lengths?*

- Switch places with the volunteer and have them repeat the activity so that now you will time and the volunteer will jump rope using the three different jump rope lengths. *Was the same jump rope length the "best" one for both of you or did each person do better with different lengths? Was the same rope length the "worst" one for both of you?*

EXTRA

Do this activity again, but this time, count both the successful jumps and the number of times the jumper messes up. *Did the rope length affect the number of mess-ups?*

SCIENCE FAIR IDEA

Listening to music may affect a person's jumping ability. Try jumping rope to slow music, fast music, and no music. *Does the music change how many successful jumps you can make in a minute? How about the number of successful jumps you can make in a row without messing up?*

SCIENCE FAIR IDEA

Can jumping rope help you on a spelling test? Randomly assign volunteers to two groups. One group will not jump rope, but will copy down 10 words from a spelling list with pen and paper. Members of the other group will jump rope and work with a partner who will call out each word and its spelling first; the jumper will then repeat it, "jumping out" each letter of the word—one jump per letter. *How well do the volunteers from each group spell the words the next day (using pen and paper)?*

OBSERVATIONS AND RESULTS

Were the most jumps in one minute made using the medium- or short-length jump rope? Were the fewest jumps made using the longest?

Using the longest jump rope should have resulted in the fewest number of jumps per minute, whereas using the short- or medium-length ropes should have resulted in the greatest number of jumps per minute, depending on how skilled the jumper is. Not only does the longest rope take the longest amount of time to turn in a full circle, you may also have found the longest one to be a bit unwieldy. The medium-length jump rope is a good length for jumping and may lead to the smallest number of mistakes. The shortest length turns the fastest, but jumpers have to be skilled to use it well and not make more mistakes with it (often by catching the rope on their heads or feet) than they did using the medium-length rope.

CLEANUP

Return all materials to where you found them.

Make Your Own Sports Equipment

DESIGN, BUILD, SCORE! CAN YOU MAKE YOUR OWN SPORTS EQUIPMENT THAT STANDS UP TO THE TEST OF REAL-WORLD PHYSICS?

Do you play or watch a sport where you use a tool to move a ball? Golf, baseball, tennis—there are many different sports where players use something other than their feet or hands to move a ball. In this activity, you will design, build, and test your own sports equipment using recycled materials.

PROJECT TIME

6 to 8 hours

KEY CONCEPTS

Engineering
Physics
Force
Speed
Design

BACKGROUND

How many sports can you think of where the athletes use something to hit a ball or a puck? Think about the size and shape of that item used in each sport. You'll quickly realize that they are all very different. A baseball bat is long and skinny, for example, whereas a tennis racket is short and wide. Golf clubs and ice hockey sticks have long handles, but the head of a golf club is much smaller than the blade of a hockey stick. What do you think would happen if you tried to play hockey with a tennis racket or baseball with a golf club? It probably wouldn't go very well!

Now think about the requirements for the equipment in each sport. They are all used to hit a ball (or puck), applying a force and changing the ball's speed or direction but in very different situations. Golf players hit a small ball that is sitting stationary on the ground, so they can do so accurately using a club with a small, heavy head while standing still. They need to hit the ball very far (hundreds of yards), which requires a powerful two-handed swing. Tennis players, on the other hand, need to move around quickly as they hit a moving ball back and forth across a net. Sometimes the ball is traveling more than 100 miles (161 km) per hour. The wide face of the racquet makes it easier to hit the moving ball (imagine trying to do it if the racket was the size of a golf club's head)—and using only one hand allows for more agility.

Now imagine that you want to play a game with your friends—but you don't have any rackets, clubs, or sticks available, so you will have to build your own! But what you need to build will depend on what type of sport you want to play. You will need to use the engineering design process to make your own sports equipment.

MATERIALS

- Assorted recyclable materials: plastic bottles, cardboard boxes and tubes, and so on.
- Duct tape and/or glue
- Scissors
- Pencil and paper
- Lightweight plastic or rubber ball (Ping-Pong ball, Wiffle ball, and so on). Do not use a heavy ball, such as a baseball.
- Open space, such as a large room or flat space outdoors
- Equipment (such as chairs and/or a table) you can use in your game
- A few friends to play your game with you (optional)

PREPARATION

- Decide what sport you want to play, and set up an area where you can play it. For example:
    - You could play floor hockey by clearing some space in the middle of a large room and putting two chairs at opposite sides of the room. The legs of the chairs can form the goals.
    - You could play mini golf by cutting out circular pieces of paper for "holes" and putting them on the floor, then setting up obstacles to form a golf course.
    - You could play table tennis by using some cardboard to make a "net" to go across the middle of a rectangular table.
    - You could also make up your own sport or use an imaginary sport, such as Quidditch from the Harry Potter series.

PROCEDURE

- You might be tempted to start building something right away, but engineers must work more systematically. First you will think through what you need to build and draw some sketches.
- Think about the sport you want to play, and what the real equipment looks like for that sport—or what you want the equipment to look like for an imaginary sport.
- Make a list of criteria for the equipment. *For example, will you hold it with two hands or one? Will it need to hit a moving ball or a stationary one? Will the ball be in the air or on the ground?*

- Look at the materials you have available. Think about how you could combine or connect them to make the equipment. *What items could you use for the handle? What about the part that hits the ball? Will you need to cut or fold them? How will you connect them?*

- Draw a sketch of your sports equipment and label the materials you will use.

- Now it's time to build! Try to build your sports equipment following the design in your sketch. You don't need to stick to your sketch exactly. You might discover that things don't work as planned. For example, maybe two pieces don't fit together like you thought they would. This is OK! You can go back to the drawing board if you need to make a change.

- When you are done building, it's time to test. Try hitting the ball a few times (if necessary, depending on your sport, get a friend to throw or hit the ball to you). *Can you hit the ball accurately? What about distance? Is your device easy to handle? Is it sturdy or is it too flimsy?*

- Based on your observations, decide if you need to make any changes or improvements to your device. You might even decide to start over and build a new one. Again, this is OK! This process is called "iteration." Engineers usually go through multiple iterations before arriving at a final design.

- Once you have arrived at a final design, build more copies of the device so you can play with your friends.

OBSERVATIONS AND RESULTS

You probably discovered that you could build a simple device to hit a lightweight ball. For example, you can make a golf club or hockey stick by taping together paper towel tubes to form a handle and a plastic bottle to form the head. Your homemade sports equipment, however, was probably not as sturdy as the real thing. Most baseball bats, hockey sticks, and so on, are made from a single, solid piece of wood or metal. You probably had to use duct tape or glue to connect multiple pieces: for example, taping multiple paper towel tubes together to make a handle. These connections introduce weak spots. Cardboard is also much flimsier than solid wood or the superstrong materials such as carbon fiber used to make tennis rackets. So while you can have fun playing your game, you probably need to be careful—your equipment might not hold to too much abuse.

CLEANUP

Recycle any materials that you can. Return all other materials to where you got them.

Why Are Skis So Long?

HIT THE SLOPES (INDOORS)—AND USE PHYSICS TO LEARN ABOUT OPTIMAL SKI DESIGN!

While watching the Winter Olympics or other winter sporting events, you might have wondered why skis are so long. The skis used by professionals are of top-notch quality. Researchers tinker with every detail, from the materials they are made of to the shape into which they are cut. The equipment is tested carefully because even the smallest detail can give an athlete the leading edge on the slope. But why would you put planks under your shoes in the first place? Do this activity to find out!

PROJECT TIME

60 to 90 minutes

KEY CONCEPTS

Physics
Weight
Pressure
Force
Center of mass

BACKGROUND

Earth's gravity pulls us all the time; it makes us return to the ground when we jump, and it makes objects fall when we release them in midair. But gravity even pulls us down when we are standing on the floor! You see this when you sit on a mattress: your weight, or how much Earth pulls on you, pushes the mattress down. As a result, the mattress squeezes together where you sit. If you stand on the mattress, you should notice it compresses more than if you lie down or sit on it. When you are standing, all your weight pushes the small area where your feet touch the mattress. If you sit or lie down, your weight is distributed over a larger area, so for each square inch that your body is touching the mattress, there is a much smaller force, and the mattress does not become as compressed.

Snowshoes use this concept to make you sink less into fresh, fluffy snow. Snowshoes create a larger contact area with the snow so your weight pushes less on each square inch of it. The strength of each push per square inch of contact area is what scientists call pressure, and snowshoes reduce that pressure.

A larger contact area with the snow or floor has more advantages. Gravity acts on us as if all our bodies' mass is pushed into a single point, and gravity pulls on that point, which is called the center of mass. If this point is located above the contact area we make with the floor, we will not fall over. The larger the contact area, the easier it is to keep our center of mass above it, and the more stable we will feel.

MATERIALS

- Baking pan or other container with sides at least 1 inch (2.5 cm) high
- Flour
- Action figure or doll that can stand upright, preferably a heavier one (and permission to temporarily glue skis on its feet)
- Cardboard
- Scissors
- School glue

PREPARATION

- To make skis for your action figure, cut two identical rectangles out of cardboard. The length should be a little shorter than the height of the figure.
- Pour about a 3/4-inch (1.9 cm) deep layer of flour onto the plate. Carefully sweep the top layer with the edge of a piece of cardboard to create a flat surface.

PROCEDURE

- Imagine the flour is fluffy snow. *What would happen if you stepped onto this fluffy snow? Will the same happen when you place your action figure there?*
- Place your action figure without skis on the flour. Pick it back up and observe. *Was your prediction correct? Do you see footprints in the flour?*
- *Do you think it would be easy to make your action figure sink into the flour? Why?*
- Place your figure back on the flour—but this time, press it down. *Was your prediction correct?*
- Pick up your action figure and let it stand on the flour without pushing it in. Try to knock it over to the side, then the front and the back. *Is it easy to make your action figure fall?*
- Dust off the feet of your action figure, then glue the skis onto its feet about halfway along the length of the skis. Let the glue dry.
- *What do you think will be different if we use the action figure with skis for the tests we did before?*
- Smooth the flour again with the edge of a piece of cardboard.

- *What type of footprints does the action figure with skis create? Are they deeper or shallower compared with the figure without skis?*
- *Is it easier or harder to make this action figure sink into the flour? Why would that be beneficial when you want to slide over the snow?*
- *Can you knock over the action figure with skis as easily? Is knocking it so it falls to the front or the back as easy as making it fall to the side? Why would this be so?*

EXTRA

Freeze a plastic tray of water to make a frozen pond and see what happens if your action figure stands on ice. *Does it leave footprints on the ice? Do people sink into ice like they sink into the snow? Could you still benefit from wearing skis when walking on ice?*

EXTRA

Look up the characteristics of Arctic animals. *Do they have bigger paws than similar animals in warmer climates? If so, why do you think that is?*

EXTRA

Give your action figure ice skates. To make one ice skate out of cardboard, you will need two small rectangles a little longer and wider than your action figure's foot. One rectangle represents the shoe and will be glued to the bottom of the foot of your action figure. The other rectangle represents the blade. Attach it vertically to the shoe (the first rectangle). This makes one skate. Repeat for the second skate. *What impressions do skates make on the flour and the ice? Can you easily knock over an action figure balancing on skates? Why do you think this is the case? Why would athletes competing on snow wear skis, whereas athletes competing on ice would wear skates?*

OBSERVATIONS AND RESULTS

Were the action figure's footprints shallower when it was wearing skis? Was it harder to push your action figure with skis into the flour, and was it harder to knock it over to the front and the back compared with when the action figure wasn't wearing the skis? This is expected.

Skis create a large contact area between the skier and the snow. Therefore, the weight of the skier—or how much gravity pulls on them—is spread out over a larger area. The skier presses less on each square inch of snow, and instead of sinking into it he or she can glide over it. You could also feel this when you pushed your action figure into the flour. Without the skis, it was easy. With the skis, your force was distributed over a large area. It was more difficult to make the action figure sink into the flour.

Skis also help with stability. It is easier to keep your balance when you have a bigger contact area. Think about how you tend to fall forward or backward when you stand on your toes; standing on flat feet is easier. Similarly, skis extend the feet, making it hard to fall forward or backward, but as they barely extend your feet to the side, skiers can still easily fall sideways.

Athletes competing on ice, however, use skates because they want their weight concentrated on a small area.

CLEANUP

Remove the skis from your action figure's feet, and peel off any remaining glue. Dispose of the flour in the compost or trash.

Energetic Two-Ball Bounces

BLAST OFF! HARNESS THE ENERGY OF TWO BALLS BOUNCING AT ONCE TO SEND ONE FLYING IN THIS SIMPLE PHYSICS ACTIVITY.

How many ball sports can you name? How many of those have several balls at once in the game? Almost none, right? Games that do use several balls at a time most likely use balls of the same mass, volume, and material. Would having two balls of different masses make a game very difficult? In this activity you will explore what might happen if you were to add a tennis ball to a basketball game or a tiny Ping-Pong ball to a tennis game or any other combination. Ready to be surprised? Try it out!

PROJECT TIME

30 minutes

KEY CONCEPTS

Physics
Energy
Collisions

BACKGROUND

Balls used in sports come in all sizes, masses, and materials. A lot of them bounce. Whether it is a basketball bouncing on the ground, a tennis ball bouncing off a racket, or pool balls bouncing off one another, the bounce can be seen as a collision. Let's see if physics can explain what happens.

When a bouncing ball falls, it initially gains speed or kinetic energy—the energy of motion. When it reaches Earth, it collides head-on with an incredibly massive object that is, from your perspective, at rest. The ball slows down, deforms temporarily, and shoots back up. The air in the ball acts like a spring—it gets compressed and expands again. During the collision, some of the ball's energy is converted into heat. As a consequence, the ball shoots up with less energy than it had when it reached Earth. Our planet, being so massive, does not move as a result of the collision.

It is interesting to compare a heavy and a light ball as they fall from the same height. Both balls will fall at a similar speed, but because kinetic energy is proportional to the mass of the object, the heavy ball reaches Earth with more energy. It will not necessarily rebound higher, as it also needs more kinetic energy to reach a specific height again. What if we could give the kinetic energy of the heavy ball to the lighter ball? When two balls collide, they exchange energy. Can we let one ball fly off with the energy of the other—and if so, how? Do this activity to find out!

MATERIALS

- A basketball or other large bouncing ball
- A tennis ball or other bouncing ball that is smaller and lighter than the bigger ball
- A Ping-Pong ball, smaller and lighter than both other balls (optional)
- A hard surface to bounce balls, preferably outdoors, surrounded by a large open area where a ball can fly without colliding with people or objects
- A helper (optional)
- Camera that films video (optional)
- Yardstick that measures meters (optional)

PREPARATION

- Gather your test balls in the open area where you will perform the activity.
- **Note:** If your ball does not bounce well, it might be deflated or old. You can add air to some balls like basketballs, soccer balls, volleyballs, etcetera. A bicycle pump with a special piece to inflate balls can help you do the job. Other balls, such as tennis balls or baseballs, wear out with use. *Why do you think a deflated ball does not bounce well? Why would a well-used tennis ball not bounce as well?*

PROCEDURE

- First, hold the larger ball in front of you at shoulder height and let it fall without giving it a push. *How high does it bounce back—almost up to your shoulder, your hip, your knee?*
- Repeat this a couple more times. *Does the ball bounce back to approximately the same height each time?*
- Repeat the test with the smaller ball. *Do you think it will bounce higher or lower then the larger ball? How high does this ball bounce back?*
- Repeat a couple more times. *Are your results consistent?*
- *Do you think the ball can bounce back higher than the height at which you released it?*
- Hold the larger ball in front of you at shoulder height like you did initially. Place the smaller ball on top of the larger ball so they just touch and hold it there. You might need a helper to hold the balls in place. In a moment, you will release the two balls at the same time. *What do you think will happen? Will the larger fall faster or will both stay together as they fall? How will they bounce back? Will a ball bounce higher, the same or lower than before?*

- Try it out and observe. You might need to try a couple of times to get the timing right. *Was your prediction correct? Can you explain what you observe?*

- Repeat the previous step a couple more times. *Are the results consistent?*

- Compare the balls' masses. *Was the top ball (the smaller one) heavier or lighter than the bottom ball?*

- Hold the smaller ball in front of you at shoulder height like you did initially and place the larger ball on top. *What do you think will happen when you release both balls at the same time?*

- Perform the test a few times. *Was your prediction correct? Can you explain your observations?* Note that although before the larger bottom ball was probably heavier than the smaller top one, this time the top ball is heavier.

EXTRA

If you have more balls available, try other combinations such as a Ping-Pong ball on a basketball or a tennis ball. If you can, try stacking three balls, such as a Ping-Pong ball on top of a tennis ball that is resting on a basketball, and release all at the same time. Be sure you have lots of free space for the balls to fly!

EXTRA

If you would like a detailed view of what happens, you can use a camera to film the experiment. Later, you can study the moving balls on a screen in slow motion. Note that you can calculate the speed (in meters per second) at which a ball travels in the video. To do so, multiply the number of frames per second by the distance (in meters) the ball traveled between two frames. Place a meter stick in your frame or use your height as a calibration of distance in your video.

OBSERVATIONS AND RESULTS

Did you find that a single ball never bounced back to the height at which you released it, regardless of the ball you used? During a collision, some of the ball's energy is converted into heat. As no energy is added to the ball, the ball bounces back with less kinetic energy and cannot reach quite the same height. Had you given the ball an initial push, you would have added energy, and the ball might have bounced back higher.

A well-inflated ball bounces better because it has more air inside. This allows it to push back faster, reducing the contact time and contact area in a collision and thus reducing the heat produced. Tennis balls also have air inside, but they cannot be reinflated.

Did you also see how a lighter ball shoots high into the air when released at the same time on top of a heavier ball? Both balls fall at the same speed, but the heavier ball gains more energy during the fall. When the lighter ball bounces on the heavy ball, they exchange energy, and the lighter ball flies off with some of the energy of a heavier ball. It reaches way higher than the height from which it was released. The heavy ball, on the other hand, is left behind with little energy and does not move much.

CLEANUP

Return all materials to where you found them.

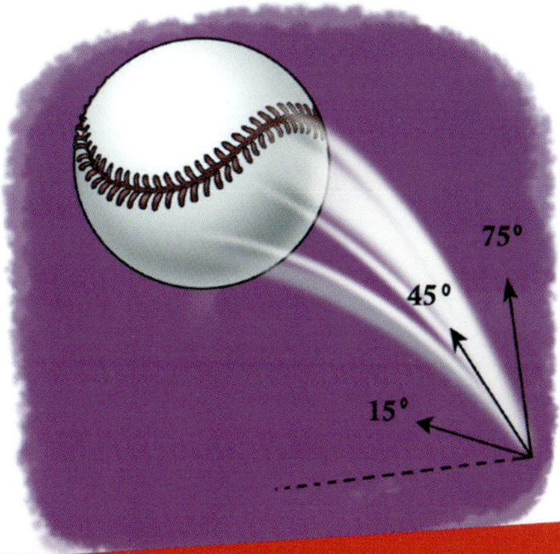

The Physics of Baseball
How Far Can You Throw?

DOES THE ANGLE OF YOUR THROW CHANGE HOW FAR A BALL TRAVELS? GIVE THIS EXPERIMENT A TRY TO FIND OUT!

PROJECT TIME

1 hour

KEY CONCEPTS

Physics
Motion
Gravity
Resistance
Angles

Have you ever seen a player make a long, game-winning pass in a football game, a baseball player throw in a ball from way out in the outfield, or a soccer goalie punt a ball all the way down the field? All of these actions involve a player trying to throw or kick a ball as far as possible. There is some science behind these throws and kicks. And sometimes it is all about the angle. The angle can affect just how far a throw or a kick can carry a ball. Should you make a shallow throw (straight out, parallel to the ground), a steep throw (throwing the ball high up in the air), or something in between? Try this sports science project to find out!

BACKGROUND

An object that is thrown, kicked, or otherwise launched through the air is called a projectile. The study of how projectiles move through the air is called projectile motion. When a projectile is launched, it has an initial velocity (its speed and direction of motion). When a projectile is moving through the air, however, it is subject to the force of gravity, which causes it to move down toward Earth. It is also subject to the force of air resistance, which slows the projectile down.

When air resistance is very low, a projectile moves in the shape of a parabola—a type of mathematical curve. (You can do an internet search for "parabola" to see what one looks like.) The path the ball follows (called its trajectory) will determine how far it goes before it hits the ground. Adjusting the ball's initial angle and keeping its initial speed constant is one way to find out the best angle to throw a ball as far as possible. In this experiment you will throw a ball as hard as you can while changing the ball's initial angle relative to the ground. What angle do you think will make the ball go the farthest?

MATERIALS

- Sports ball of your choice (baseball, tennis ball, football, etcetera)
- Large open area to throw the ball
- Three each of three different objects to mark where the ball lands on the ground (for example, three sticks, three rocks, and three golf balls)
- Helper

PREPARATION

- Make sure you have a large, open area to throw and that there are no other people or objects in the way.
- Do some stretching or warm-up exercises before you start the experiment—it will involve a lot of throwing!

- Pick a place to stand in your large, open area. Mark a place on the ground so you know to always throw the ball from that location.
- Pick a direction to throw your ball. (If it is a windy day, make sure you pick a consistent direction relative to the wind.)
- *What type of angle do you think will get the farthest distance—shallow, steep, or medium?*

PROCEDURE

- Throw your ball as hard as you can, relatively straight forward at a "shallow" angle—as parallel to the ground as you can. If you are familiar with measuring angles in degrees, imagine trying to throw the ball at about a 15-degree angle relative to the ground. *Can you think of a time this type of throw would be best?*
- Have your helper mark where the ball lands (for example, with a rock).
- Repeat the throw at that angle two more times. Try to consistently throw the ball as hard as you can, so its initial speed is constant. Have your volunteer mark each throw. *Why do you think it's important to perform the same test at least a few times?*
- Now repeat the procedure for a "medium" angle (outward and upward) of about 45 degrees three times. Remember to try and consistently throw the ball as hard as possible. Use a different marker object for where these throws land. *How did this distance compare with the shallow angle?*

- Now, do the same for a "steep" angle (up in the air and slightly forward but not straight up) of about 75 degrees. *How do you think this throw distance will compare with the others?* Remember to try and consistently throw the ball as hard as possible. Use another different type of marker for this batch of throws.

- Now, analyze the results of your experiment by looking at the markers. *Which throws went farthest? What angle do you think is best if you want to throw the ball as far as possible?*

SCIENCE FAIR IDEA

Try the experiment using different types of balls. *How do you think the ball size, weight, and shape will affect what angle is best to throw the farthest? Did you get different results?*

EXTRA

Repeat the exercise, this time kicking a ball instead. *Do you think the angles will yield relatively similar distances? Did you get different results than when you threw the ball?*

OBSERVATIONS AND RESULTS

In most cases, you should see that a medium, 45-degree angle made the ball go the farthest. When you throw the ball at a very shallow angle, gravity pulls it down and the ground is closer, which ends the throw sooner, so it does not have as much time to travel as far forward. When you throw the ball at a very steep angle, most of its velocity is directed upward instead of forward, so it does not make much forward progress.

Depending on the type of ball and the amount of air resistance on it, however, you might see different results. A ball that has a lot of air resistance (such as a Whiffle ball) might slow down significantly when it is in the air for a long time, so it might actually go farther at a shallower angle.

CLEANUP

Return all materials to where you found them.

Surface Science

Where Does a Basketball Bounce Best?

CAN THE FLOOR'S MATERIAL CHANGE HOW HARD A PLAYER NEEDS TO DRIBBLE A BASKETBALL? FIND OUT IN THIS ACTIVITY

PROJECT TIME

1 to 2 hours

KEY CONCEPTS

Energy
Gravity
Physics
Sports

Playing basketball can be hard work. Players not only have to run around the court but also dribble the basketball at times. That takes some serious effort too. Have you ever wondered why that is? The challenge has to do with how the basketball bounces. When the ball hits the court, its bounce actually loses momentum by transferring some of its energy into a different form. This means that to keep the ball bouncing to the same height, players must continually put energy into the ball with each bounce. In this activity, you will explore how high a basketball bounces on different surfaces compared with the height from which it was dropped. Which surface lets the basketball bounce the highest? Grab a basketball and try this activity to find out!

BACKGROUND

Playing a game of basketball can be great exercise, and one part of that workout comes from just dribbling the ball. Why is this? When a basketball hits the ground (and as it flies through the air), it actually transforms some of its energy to another form. If players do not put enough energy back into the ball, they will not be able to dribble it effectively.

When a basketball bounces, it has two different types of energy: kinetic and potential. Kinetic is the energy an object has due to its motion. Potential energy is that which is stored in an object—its potential for motion—such as due to its height above the ground. For example, when you hold a basketball at waist level, it has some potential energy. If you drop the basketball, the force of gravity pulls it down, and as the ball falls, its potential energy is converted to kinetic energy. When a basketball hits a court floor, a part of the kinetic energy gets converted into sound or heat, some of it briefly changes the ball's shape (flattening it slightly) and a portion is absorbed by the floor surface.

MATERIALS

- At least two different surfaces to bounce a basketball on, with at least one hard surface and one soft surface (For example, you could use carpet, concrete, grass, linoleum, and a basketball court. The surface needs to be flat and next to a wall or another large perpendicular surface.)
- Tape measure or yardstick
- Painter's tape or masking tape
- Basketball
- A helper
- Video camera and access to a computer or a large screen on which to watch your recorded video (optional)

PREPARATION

- Prepare the walls or other vertical surfaces next to the floor types you want to test so that you can estimate the height of the basketball's bounce. To do this, use a tape measure or yardstick along with the painter's or masking tape to mark every 8 inches (20 cm), starting from where the wall meets the floor and going up to 40 inches (102 cm) high on the wall. You should end up with five tape marks on each wall.
- If you are using a video camera, ask your volunteer to set it up so that all of the marked wall measurements as well as the floor are in view. When you're ready to test the basketball on a surface, ask your volunteer to start the video camera.
- If you are not using a video camera, ask your volunteer to get ready to watch when you drop the basketball to see roughly how high it bounces after it first hits the ground.

PROCEDURE

- Hold the basketball so that the bottom of it is lined up with the top edge of the highest tape mark you made.
- Drop the ball. (Do not push it down.)
- Let the basketball bounce back up and then hit the ground a second time before you catch it in your hands. (Then stop recording if you were doing so.) *How high does your volunteer say the basketball bounced after hitting the ground the first time?*
- Repeat this, dropping the basketball on the same surface a few more times to give you a good idea of just how high it bounces when being dropped from a certain height.
- Repeat this entire process with the other surface(s) you want to test. *How high does the basketball bounce off another surface compared with the first one you tested? Why do you think this is?*

- **Tip:** If you are testing a surface that is at a very different temperature (such as concrete outside on a cold day), you will want to do your testing quickly so that the ball does not change temperature too much. A change in the ball's temperature can also affect how it bounces.
- If you videotaped the basketball bounces, watch your videos to try to more closely estimate the basketball's bounce height on the different surfaces.

EXTRA

Try to quantify your results from this activity. To do this, you'll want to record your bounce trials and closely watch the videos on a large screen to determine the exact height of the basketball before it was dropped and the highest point of its first bounce. You could even graph your results. *Exactly how high does the basketball bounce on each of the different surfaces?*

SCIENCE FAIR IDEA

A basketball loses kinetic energy by transferring it into other forms when the ball bounces. *But just how many bounces can a basketball make before it loses all its kinetic energy and stops bouncing? And how does this change if you alter some factors, such as the type of surface the basketball bounces on or the drop height?* Design an experiment to investigate how many bounces a basketball can make and how various factors affect that number, and then try it out!

OBSERVATIONS AND RESULTS

Did the basketball bounce much higher on the harder surface compared with the softer one?

One factor that can affect the basketball's collision with the ground is the type of surface the ball collides with. When a basketball bounces off of a surface, some of its energy is absorbed by that surface. Some surfaces absorb more energy than others do. (How much energy gets absorbed determines how much energy a player has to put back into the ball to keep it bouncing.) A hard surface, such as concrete, absorbs less energy compared with a soft surface, such as a carpeted floor. The more energy absorbed by the surface, the less that remains in the ball for it to bounce. This is why you should have seen that when you bounced the basketball on a relatively hard surface, it bounced higher (it lost less energy) compared with when it was bounced on a softer surface (where it lost more energy). For example, depending on the type of basketball and surface, you may have seen the ball bounce about 15 inches (38 cm) high on carpet and about 25 inches (63.5 cm) high on concrete.

CLEANUP

Remove tape from the wall or other perpendicular surface. Return all other materials to where you found them.

Swiveling Science

Applying Physics to Hula-Hooping

WILL THE WEIGHT OF A HULA-HOOP CHANGE HOW IT SPINS? GIVE THIS ACTIVITY A TRY TO SEE!

Have you spent time this summer spinning a Hula-Hoop around your waist or arm? Could you do it easily or was it difficult? Have you ever wondered how Hula-Hoops work or what makes them able to spin around a person's waist or arm, seeming to defy gravity? The answer can be explained by physics, which can help you determine what makes an effective hoop. In this activity you'll get to create your own hoops and investigate how their weights affect how they spin. Which do you think will spin better, a heavy hoop or a lighter one? Get ready to do some hooping to find out!

PROJECT TIME

1 to 2 days

KEY CONCEPTS

Physics
Speed
Forces
Weight
Gravity
Geometry

BACKGROUND

What makes a Hula-Hoop spin around a person's waist? It comes down to a combination of several factors. Every object has weight, which pulls it down toward the ground. In order to lift something off the ground, you need to exert an upward force (a push or a pull). In order to make something spin, you need to exert a push or pull that creates torque (such as when you twist a doorknob). So, to make a Hula-Hoop spin and stay up in the air, you need to exert an upward force and a torque with your hips. Another force involved in the hula-hooping process is friction, the force resisting two surfaces sliding against each other. For example, if you push a book and let it slide across a table, it will slow to a stop due to friction. If you want it to keep moving, you have to keep pushing it. Similarly, friction between the hoop and the Hula-Hooper's clothes and the air will cause the hoop's spinning to slow down. To keep it in the air you have to keep spinning it and pushing it up. Friction, however, also helps prevent the hoop from sliding downward when it's in contact with your hips.

That might sound like a lot of complicated physics for a toy, but don't worry! You will do some hula-hooping in this project and explore for yourself. How do you think the hoop's weight affects how it spins?

MATERIALS

- Tape measure
- Calculator
- PVC cutter (preferred), or a sharp cutter knife, or some other type of cutter that can safely cut through polypipe
- Polypipe, which is hard black tubing usually used for irrigation, about 25 feet (75 m) in length (5/8 inch [1.6 cm], 3/4 inch [1.9 cm], or 1 inch [2.5 cm] diameter) Available from a garden supply or home improvement store.
- Poly insert coupling or wooden dowel that fits snugly inside the polypipe tubing. For example, if you are using 3/4-inch (1.9 cm) tubing, you will want to use a 3/4-inch (1.9 cm) poly insert coupling. This should also be available from a garden supply or home improvement store.
- Hair dryer
- Wide plastic tape such as duct tape
- One cup of sand
- Measuring cup

- Funnel with an opening that can fit into the polypipe or a thick sheet of paper and tape
- Bucket (optional)
- Timer or stopwatch
- Adult helper
- Another helper to assist in testing the hoops (optional)

PREPARATION

- Measure the height from the ground to somewhere between the Hula-Hooper's navel and the middle of their chest. This will be the height (or diameter) of the hoop.

- To figure out the length of polypipe you want to cut to make your hoop, take the number you just measured and multiply it by pi (3.14). (The circumference of a circle equals pi times the circle's diameter.)

- Measure the length of polypipe needed and have an adult carefully cut the tubing at the right spot using a PVC cutter or sharp cutter knife.

- Use a poly insert coupling (or wooden dowel) to connect the ends of the tubing so they form a circle. If it's a tight fit, use the hair dryer to warm the tubing ends (one at a time) for about two minutes before inserting the coupling inside the tubing. Insert the coupling in one end of the tubing until the coupling is about halfway inserted, then insert the other half of the coupling into the tubing's other free end. Push the tubing together until little (or no) coupling is visible, using the hair dryer to heat the tubing if it is difficult to insert the coupling.

- Put a short strip of wide plastic tape over the connection where the two ends of tubing meet.

- Follow this process to make a second hoop that is the same size, but this time add one cup of sand to the tube (to increase its weight) before connecting the ends with the poly insert coupling. To add the sand, use a funnel to carefully pour the sand into the tubing. You may want to do this over a bucket and/or have a helper assist with this. (You should pour any spilled sand into the tubing.) If you do not have a funnel, you can make one by rolling a thick sheet of paper into a cone shape with a hole at the end, fastening it with tape. *When you're done making both hoops, how do they feel compared with one another?*

PROCEDURE

- Now you will do some hula-hooping with your homemade hoops. If you need directions or tips on how to hula-hoop, you can watch an instructional video onlin.

- Pick up the lighter two hoops to try first.

- Have one person (yourself or a helper) hula-hoop with it around their waist while a helper starts timing the Hula-Hooper as soon as they reach a steady pace. Have the helper time the person for one minute and count how many full turns the hoop makes during that time. If the Hula-Hooper cannot go through the full minute with the hoop moving, try starting over again or try collecting data for only 30 seconds. *How many times could the Hula-Hooper spin the lighter hoop in a minute? How well could they spin it around? Does it seem awkward or does it spin well?*

- Repeat this process with the heavier hoop, again timing how many spins the Hula-Hooper can do in one minute. (Make sure the hooper doesn't change clothes while collecting data.) *Did the Hula-Hooper spin the heavier hoop faster or slower than the lighter one? Why do you think this is? Did this hoop feel more or less awkward to spin, or about the same as the other hoop?*

- *Which hoop spun faster, the lighter or heavier one? Was the faster hoop also the one that was the easiest to spin, once it got going?*
- If you want, you can repeat this process a few more times for each hoop. *Are your results consistent?*
- For even more data, the Hula-Hooper and helper can switch roles and repeat this process for both hoops. *Did the other person get similar results?*

SCIENCE FAIR IDEA

In this activity, you compared hoops that were the same size but had different weights. You could repeat this activity again, but this time make hoops of different sizes. Tip: The smallest hoop should have a diameter up to the Hula-Hooper's navel, and the largest should have a diameter to the middle of their chest. *How does the size of the hoop affect how it spins?*

SCIENCE FAIR IDEA

You could make a broader range of oops, such as by adding different amounts of sand and/or making several different sizes. *Is there a relationship between speed, the weight and/or size of the hoop? If so, what is it?*

SCIENCE FAIR IDEA

Try hula-hooping in different types of clothes. *Do different garments affect the speed at which the hoop rotates? Can you correlate your results to the force of friction?*

OBSERVATIONS AND RESULTS

Did the heavier hoop spin slower than the lighter one? Was it harder to spin the heavier hoop for a long time than it was to spin the lighter one?

The hoop with sand added to it should have clearly felt heavier than the hoop that held no sand. Because of its greater weight, the heavier hoop was pulled down more as it was spun around in the air. The experimenter probably felt a need to work harder to keep the heavier hoop up and spinning, and it might have even been difficult to keep it at waist level for more than 30 seconds. The heavier hoop probably spun much slower than the lighter one (such as around 60 to 70 turns per minute for the heavier hoop compared with 100 to 120 turns per minute for the lighter one), although there can be a lot of variability, depending on the Hula-Hooper and the hoops.

CLEANUP

Store the hoops to use again, if you'd like.
Return all other materials to where you found them.

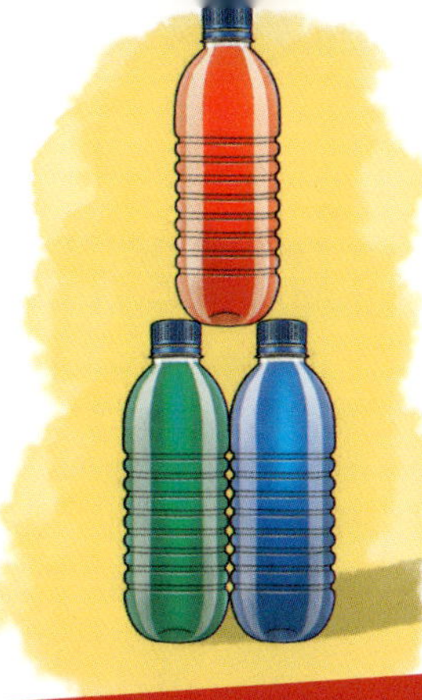

Sporty Science

The Mechanics of a Carnival Game

IS THERE SCIENCE INVOLVED IN CARNIVAL GAMES? THE ANSWER MIGHT SURPRISE YOU!

Have you ever played a game at a carnival, trying to win a stuffed animal or another prize? It might look easy—until you try it. Why are those "simple" games at fairs, carnivals, and other festivals so hard? Is it really lack of skill or coordination, or do midway vendors use some basic laws of science to help them set up the games in their favor? In this science activity, you'll investigate how physics can help you win—or lose—at the classic game of trying to knock over a pyramid of milk bottles using a ball.

PROJECT TIME

1 to 2 hours

KEY CONCEPTS

Mass
Center of mass
Balance
Physics
Games

BACKGROUND

Why can it be nearly impossible to knock down pins or hit the right target to win that giant stuffed animal at the carnival or fair—especially when it looks so easy? To answer this, in this activity we'll look at the classic carnival game sometimes called "One Ball," in which milk bottles are stacked in a pyramid and you get one throw with a ball to try to knock them all over. To beat this game, it's useful to think about how redistributing an object's mass can affect how well it balances. For example, it might be easy to stand on a balance beam while holding a heavy backpack hanging down in front of you, but it's much more challenging when that backpack is on your back.

How an object's mass is distributed can affect its center of mass, which is the average location of most of an object's mass. This basically means that the center of mass will be in the object's center if the mass is evenly distributed on top and bottom (as it usually is with a ball). And it will shift to the object's heavier side if the mass isn't uniformly distributed. So, for a pyramid, the center of mass will be much lower than its physical center because so much more of the mass is located in the lower half.

MATERIALS

- A very large room or an outside area (You will need plenty of space so you can throw a ball without hitting anyone or anything.)
- A small, stable table
- Masking tape, a stick, a rock, or a similar object to mark off a throwing distance
- Tennis ball or baseball
- Three plastic bottles filled with water, all the same shape and size (Make sure that you can stably stack the bottles in a pyramid shape. Most 16.9-fluid ounce [0.5 l] drinking-water bottles should work well for this.)
- Food coloring (optional)

PREPARATION

- Fill each of the bottles with water. The same amount of water should be in each bottle.

- If you want, you can remove the labels from your bottles and add some food coloring (three drops) to each bottle to give your carnival game some color. If you add food coloring to the bottles, make sure you do your testing where it will not be a problem if a bottle spills some dyed water! (If you use dyed water, it's recommended you do this activity outdoors.)
- Take your materials to the very large room or area outside where you can set up your mini carnival game.
- Put the small table at a set distance from your "throw line." You might try about 8 feet (2.4 m) away from the table. Mark the throw line using a piece of masking tape, a rock, a stick, or a similar object.

PROCEDURE

- Make sure the water bottle lids are on tight. Stack the three bottles into a stable pyramid shape on the table, with two bottles on the bottom and one on top that is centered between them and resting on their lids. *How stable is your pyramid?*
- From your throw line, throw the tennis ball (or baseball) at the bottle pyramid. *Did you hit any bottles? If so, which bottle was hit? How many bottles were knocked over?*
- Arrange the pyramid as it was before on the table.
- Repeat this process until you have hit the pyramid at least 10 times with the ball. Try to throw the ball the same way each time, and try to hit each bottle a few times. *How many bottles usually get knocked over? Does it seem to depend on which bottle you hit with the ball? If so, which bottle(s) do you need to hit to knock over the most bottles?*
- Now take the top bottle from the pyramid and empty the water out of it. Stack the bottles into a stable pyramid shape on the table with the empty bottle on the top. *How stable does this pyramid seem?*

- As you did before, throw the ball from the throw line at the pyramid at least 10 times, rearranging the pyramid on the table after each throw. Again, try to throw the ball the same way each time, and try to hit each bottle a few times. *How many bottles usually get knocked over now? Does hitting a certain bottle (or bottles) tend to knock over the most bottles?*
- Take one of the bottom bottles from the pyramid and empty the water out of it. Stack the bottles into a stable pyramid shape on the table with the two empty bottles on the bottom and the water-filled bottle on the top. *How stable does this pyramid seem?*
- As you did before, throw the ball from the throw line at the pyramid at least 10 times, rearranging the pyramid on the table after each throw. Again, try to throw the ball the same way each time, and try to hit each bottle a few times. *How many bottles usually get knocked over with this pyramid arrangement? Does hitting a certain bottle (or bottles) tend to knock over the most bottles?*
- *Overall, which pyramid arrangement led to the highest number of throws where all three bottles were knocked over? In other words, which arrangement was most successful? Which was least successful? Which bottle(s) should be hit to cause the largest number of bottles to fall down?*

EXTRA

You could repeat this activity, but this time you could quantify your results. That is, when testing each pyramid arrangement, write down which bottle is hit each time and how many bottles are knocked over. *If you quantify your results, just how much more "successful" is one pyramid arrangement compared with another? How much better is it to hit one bottle compared with another?*

EXTRA

Try moving the throw line closer to the pyramid or farther away from it. *How does your throwing distance from the pyramid change how successful you are at knocking it over?*

EXTRA

Instead of bottles, you could use wooden blocks and arrange them in different configurations, such as stacking all three on top of one another. *Using wooden blocks, which configuration is easiest to knock over? Which is hardest?*

OBSERVATIONS AND RESULTS

When the pyramid was arranged with two empty bottles on the bottom (and a filled bottle on top), did it fall over the easiest? Was hitting the bottom bottles typically the best approach?

A lot of people may initially think that the center of mass of the first bottle pyramid you made is in the middle of the structure (where the upper bottle rests on the lower ones). But in this configuration, with all of the bottles filled equally, it's actually closer to the middle of the two bottom bottles—because there are two bottles on the bottom, there is more mass on the bottom of the pyramid than the top. So the center of mass is closer to the pyramid's bottom. In the second pyramid arrangement, the center of mass became even lower because more mass was in the bottom part of the pyramid compared with the top. In the third pyramid arrangement, the center of mass became much higher—somewhere within the top bottle.

You should have found that the third pyramid, with its higher center of mass, was the easiest one to knock over and most unstable, likely having all three bottles fall over when the ball touched any of them. On the other hand, the second pyramid, with its lower center of mass, was likely the hardest to knock over completely. In general, hitting the lower area between the bottom bottles (below the center of mass of the pyramid) should have been the most successful approach for knocking down the entire pyramid. Except for the third pyramid, hitting the top bottle likely only knocked the top bottle off the pyramid, leaving the two others still standing.

CLEANUP

Recycle the empty water bottles. Return all other materials to where you found them.

Speedy Science
How Fast Can You React?

HOW MUCH TIME DOES YOUR BRAIN NEED TO REACT? THIS EXPERIMENT WILL HELP YOU FIGURE THAT OUT!

Think fast! Have you ever noticed that when someone unexpectedly tosses a ball at you, you need a little time before you can move to catch it (or duck)? That's because when your eyes see an incoming signal such as a ball, your brain needs to first process what's happening—and then you can take action. In this activity, you can measure just how long it takes for you to react, and compare reaction times with your friends and family.

PROJECT TIME

45 to 60 minutes

KEY CONCEPTS

Reaction time
Neuroscience
Gravity

BACKGROUND

You may not realize it, but when your senses pick up clues from the outside world—the smell of baking cookies, the color of a stoplight, the rrrring! of an alarm clock—it takes a fraction of a second for you to recognize that signal and respond. During that time your brain receives information from your senses, identifies a possible source, and allows you to take action. The jam-packed fraction of a second is called your reaction time.

This activity teaches you about your brain's reaction time, but it also relies on the laws of physics. Specifically, you can calculate your reaction time using our handy chart, which is based on how quickly a ruler falls. How do we know how quickly your ruler will fall? Gravity pulls all objects toward Earth's center at the same speed. If you want to try this out at home, try dropping a tennis ball and a basketball from the same height: They should both hit the ground at the same time!

MATERIALS

- Ruler (inches or metric)
- Paper
- Pencil
- Chart (below)

Line Where Partner Pinched Ruler (inches | centimeters)		Reaction Time (seconds | milliseconds)	
2 in	**5 cm**	**0.1 sec**	**100 ms**
4 in	10 cm	0.14 sec	140 ms
6 in	15 cm	0.17 sec	170 ms
8 in	20 cm	0.2 sec	200 ms
10 in	25.5 cm	0.23 sec	230 ms
12 in	30.5 cm	0.25 sec	230 ms
17 in	43 cm	0.3 sec	300 ms
24 in	61 cm	0.35 sec	350 ms
31 in	79 cm	0.4 sec	400 ms
39 in	99 cm	0.45 sec	450 ms
48 in	123 cm	0.5 sec	500 ms
69 in	175 cm	0.6 sec	600 ms

PREPARATION

- You need to use some math skills in this challenge. To make things easier, we've provided a chart on the previous page that you can make a copy of. The basic rule: 100 milliseconds translates into about 2 inches, or 5 centimeters.
- On a clean sheet of paper, write the name of each person, including yourself, who will take part in this experiment. You only need two people for this activity, but it's also great for a group. Leave five spaces below each name.

PROCEDURE

- Hold the ruler vertically so that the zero end hangs down.
- Ask your partner to stand next to you and place their hand below the ruler's zero line, ready to catch the ruler when it falls by pinching it between their thumb and index finger. Your partner's fingers should be just below the ruler, but as close as possible to the bottom edge without touching or overlapping.
- Tell your partner that you will count from one to five and drop the ruler at some point during the count. Your partner will need to catch the ruler as quickly as they can, pinching the ruler between their fingers.
- Count from one to five and drop the ruler at some point.
- Your partner should catch and pinch the ruler. *How fast did your partner appear to act? Did your partner's fingers pinch near the zero line?*
- Write down the centimeter or inch line where your partner's fingers pinched the ruler.

- Calculate how long it took your partner to respond using the chart provided. *Was your partner as fast as you thought?*
- Repeat the drop four more times for your partner, and record the measurement each time. *Does your partner's reaction time change? Are the five reaction times different?* Vary when you drop the ruler: For example, you could drop on the count of five first, then drop on two.
- Switch tasks and try catching when your partner drops the ruler, then compare your results with the others. *Do most people have a similar reaction time? Are older people faster than younger people? Are girls faster than boys?*
- You can also try a few variations: *What happens when you tell your partner when you will drop the ruler? Does reaction time improve with practice?*

EXTRA

Ambidextrous, anyone? Repeat this activity and compare your results when you use your dominant hand—the hand you write with—and when you use your other hand. *Is there any difference between hands?*

EXTRA

Consider adding other distracting sounds and sights—such as turning on a TV set or flicking a flashlight on and off—during the activity. *Do your responses slow with so many sensory signals?*

OBSERVATIONS AND RESULTS

Did you and your partner usually catch the ruler at around 6 inches (15 cm)? What took so long?

On average, reaction time takes between 150 and 300 milliseconds. If that sounds like a long time, think about how much has to happen for you to react. When your eye sees the ruler falling, information travels from sensory cells called neurons from the eye to the brain's visual cortex, an area devoted to understanding what you see. Next, the motor cortex—the part of the brain that directs movement—has to send signals along your spinal cord and to your arm, hand, and finger muscles, telling them to respond in the proper sequence to catch the ruler—quick! That's a lot happening in less than half a second—and a pretty amazing feat!

CLEANUP

Return all materials to where you found them.

THE SCIENTIFIC METHOD

The scientific method helps scientists—and students—gather facts to prove whether an idea is true. Using this method, scientists come up with ideas and then test those ideas by observing facts and drawing conclusions. You can use the scientific method to develop and test your own ideas!

Question: What do you want to learn? What problem needs to be solved? Be as specific as possible.
Research: Learn more about your topic and refine your question.
Hypothesis: Form an educated guess about what you think will answer your question. This allows you to make a prediction you can test.
Experiment: Create a test to learn if your hypothesis is correct. Limit the number of variables, or elements of the experiment that could change.
Analysis: Record your observations about the progress and results of your experiment. Then analyze your data to understand what it means.
Conclusion: Review all your data. Did the results of the experiment match the prediction? If so, your hypothesis was correct. If not, your hypothesis may need to be changed.

GLOSSARY

agility: How quickly a person can move, stop, change direction, and start again.

angle: The space formed when two lines meet at a point.

collision: When two objects come in contact with each other.

cortex: The outer layer of the brain.

drag: A force that slows the movement of something through liquid or gas.

energy: The ability to do work.

force: The push or pull on an object.

friction: A force that holds back movement of objects sliding against each other.

gravity: A force that pulls objects downward toward Earth.

iteration: The repeating of a process.

mass: The amount of matter in an object.

momentum: The measurement of mass in motion.

pressure: The measure of the amount of force in one area.

speed: The rate at which an object moves over a certain distance.

torque: A twisting force that makes something rotate.

trajectory: The path of a moving object through space.

ADDITIONAL RESOURCES

Books

Spangler, Steve. *Steve Spangler's Mind-Blowing Science Experiments for Kids and Their Families!: 40+ Amazing STEM Projects You Can Do Together.* New York, NY: Topix Media Lab, 2022.

Swanson, Jennifer. *The Secret Science of Sports: The Math, Physics, and Mechanical Engineering Behind Every Grand Slam, Triple Axel, and Penalty Kick.* New York, NY: Black Dog & Leventhal Publishers, 2021.

Ventura, Marne. *Learning STEM from Basketball: Why Does a Basketball Bounce, and Other Amazing Answers for Kids.* New York, NY: Sky Pony Press, 2021.

Websites

Discovery Education
sciencefaircentral.com

Exploratorium
www.exploratorium.edu/search/science%20fair%20projects

Science Buddies
www.sciencebuddies.org/science-fair-projects/project-ideas/list

Science Fun for Everyone!
www.sciencefun.org/?s=science+fair

Videos

Sports Fitness
whyy.pbslearningmedia.org/resource/a447cf56-a61a-4539-93dc-f6a559363836/sports-fitness/

Sports Physiology
whyy.pbslearningmedia.org/resource/sports-physiology-science-trek/sports-physiology-science-trek/

Publisher's note to educators and parents: Our editors have carefully reviewed these websites to ensure that they are suitable for students. Many websites change frequently, however, and we cannot guarantee that a site's future contents will continue to meet our high standards of quality and educational value. Be advised that students should be closely supervised whenever they access the internet.

INDEX